# FLXIBLE INTERCITY BUSES

## 1924-1970 PHOTO ARCHIVE

William A. Luke

**Iconografix**
Photo Archive Series

Iconografix
1830A Hanley Road
Hudson, Wisconsin 54016 USA

© 2003 William A. Luke

All rights reserved. No part of this work may be reproduced or used in any form by any means... graphic, electronic, or mechanical, including photocopying, recording, taping, or any other information storage and retrieval system... without written permission of the publisher.

The information in this book is true and complete to the best of our knowledge. All recommendations are made without any guarantee on the part of the author or Publisher, who also disclaim any liability incurred in connection with the use of this data or specific details.

We acknowledge that certain words, such as model names and designations, mentioned herein are the property of the trademark holder. We use them for purposes of identification only. This is not an official publication.

Iconografix books are offered at a discount when sold in quantity for promotional use. Businesses or organizations seeking details should write to the Marketing Department, Iconografix, at the above address.

Library of Congress Control Number: 2003103546

ISBN-13: 978-1-58388-099-108-8
ISBN-10: 1-58388-099-108-5

Reprinted May 2012

Printed in The United States of America

Copyediting by Suzie Helberg

COVER PHOTO: In 1940, the Flxible Company introduced a modern coach in the Clipper model class. It followed a design featuring a rear engine and streamlining. The slanted side windows and stainless steel trim panels on the side of the coach were attractive design features. It is now considered that the 1940 Flxible Clipper was one of the classic coaches of the 20th Century. The Clipper pictured was purchased by Northern Transportation Company of Bemidji, Minnesota in 1940. Northern Transportation Company was typical of many of the bus companies which operated this type of coach. Also see page 43.

# BOOK PROPOSALS

Iconografix is a publishing company specializing in books for transportation enthusiasts. We publish in a number of different areas, including Automobiles, Auto Racing, Buses, Construction Equipment, Emergency Equipment, Farming Equipment, Railroads & Trucks. The Iconografix imprint is constantly growing and expanding into new subject areas.

Authors, editors, and knowledgeable enthusiasts in the field of transportation history are invited to contact the Editorial Department at Iconografix, 1830A Hanley Road, Hudson, WI 54016.

# Table of Contents

Acknowledgments ..................................................................................... 4
Introduction ............................................................................................... 5
Early Flxible Coaches................................................................................. 7
Flxible Airway Coaches ............................................................................ 11
Original Flxible Clipper Coaches ............................................................. 19
Rear-engined Flxible Clipper Coaches .................................................... 27
Rear-engined 29-passenger Flxible Clipper Coaches............................. 38
New Design Flxible Clipper Prototype ................................................... 42
New Design 25-passenger Flxible Clipper Coaches .............................. 43
New Design 29-passenger Flxible Clipper Coaches .............................. 51
Small 18-passenger Flxible Coaches ...................................................... 62
Flxible Clipper Coaches During World War II....................................... 66
Post-war Flxible Clipper Coaches ........................................................... 70
C-1 Dual-engined Flxible Clipper Coach................................................ 77
Flxible Post Office Coaches...................................................................... 78
Flxible Sightseeing Coaches .................................................................... 79
Flxible Visicoaches................................................................................... 82
Flxible Airporter Coaches ........................................................................ 93
Flxible Starliner Coaches ........................................................................ 95
Flxible Vistaliner Coaches ....................................................................... 99
Flxible Hi-level Coaches .......................................................................... 100
Flxible Flxiliner Coaches ......................................................................... 102
Flxible and Dina Coaches in Mexico ....................................................... 105
Ansair Flxible Coaches in Australia........................................................ 108
Restored Flxible Coaches ........................................................................ 110
Time Tables Featuring Flxible Coaches ................................................. 113
Flxible Sales Literature ........................................................................... 115
A Word from the Author .......................................................................... 126

# Acknowledgments

Photographs in this book are from the bus history library of the author, William A. Luke, unless noted as photo credits from other individuals and organizations.

The following persons and organizations were very helpful in providing information that has made this book possible:

Donald Coffin, Bus Industry Historian, Hawley, Pennsylvania

Tom Jones, Librarian, Motor Bus Society, Clark, New Jersey

Paul Leger, President, Bus History Association, Halifax, Nova Scotia

# Introduction

The Flxible Company's bus building played an important role in the growth and development of bus transportation in the United States between 1924 and 1960. Intercity coaches built by Flxible were seen everywhere and fit the needs of both large and small bus companies. The Flxible Company worked closely with the industry to adapt to the changes that occurred during the development years.

Hugo Young founded the Flxible Side Car Company in 1913 in Loudonville, Ohio. It was incorporated the following year. The sidecar was a unit that could be attached to a motorcycle to give a seat to a passenger. The motorcycles with Flxible sidecars were effective in World War I. They were also used in the sport of motorcycle racing. The name Flxible was chosen because of a flexible connection between the motorcycle and the sidecar that allowed the motorcycle to tilt going around curves, but kept the sidecar and its single wheel on the ground. The word "Flxible," without the middle "e," was used because the word flexible is a common-usage word and could not be trademarked.

In the early 1920s, the mass production of automobiles lowered their cost to the point where sidecar-equipped motorcycles could no longer compete. The Flxible Company used this challenge as an opportunity to enter the bus-building business. The first Flxible bus was built in 1924, and was among only a few Flxibles built in the 1920s. They were mostly stretched units on Buick chassis. The Goodyear Tire and Rubber Company purchased early Flxible buses, with one being used to pull a giant 12-foot Goodyear tire used for promotion. Another bus was equipped with a mooring mast and used to carry a ground crew and moor the blimp *Mayflower* when it was not in its hangar.

The real boost in Flxible's bus production came in 1936 with the introduction of the Flxible Airway coach, mounted on a Chevrolet truck chassis. The body first accommodated 16 passengers and later the capacity was increased to 19 passengers. The Airway coach came at an opportune time, as many bus companies were struggling in the Depression years and the Airway, with its economies and passenger appeal, helped them survive. Soon after the introduction of the Airway, the first Flxible Clipper made its debut. It was similar to the Airway model, but the body was extended over the front engine. The success of the original Clipper was followed by an entirely new model, the all-steel rear-engined Flxible Clipper. Its body style with the front entrance behind the front wheels, the rear engine position, 25-passenger capacity, and large rear baggage compartment, continued with some modifications into the 1960s.

The first rear-engined Clipper had a Chevrolet engine, and in 1940 Buick straight-eight engines were introduced in a larger, new 29-passenger Clipper. The Buick engine was an option for the 25-passenger Clippers.

More than 1,000 bus companies were reported to have purchased Flxible coaches. Many orders were for only one coach at a time, but there were some multiple orders. Most importantly, satisfaction in the Flxible coaches was evident because most orders were repeat orders. For instance, Queen City Coach Company purchased 82 Flxible coaches between 1935 and 1942. Greyhound companies bought 178 Flxible coaches.

World War II interrupted Flxible bus production. The Flxible Company was involved in building army and navy aircraft and was awarded the prestigious Army-Navy "E" award for its outstanding achievement in this endeavor. Other awards were presented to The Flxible Company for its contributions to the war effort.

Production of Flxible coaches resumed in 1944, and the Flxible Clipper models of 1941 were continued until 1946. Style changes were then incorporated, with a rounded front and curved-glass windshield becoming the new look for the Clipper.

The Flxible Company offered the coach body for other purposes such as mobile TV coaches, product display vehicles, technical and medical units, and motor homes which were referred to as Flxible "Land Cruisers."

The name Clipper was discontinued in 1950 and the new name Visicoach was adopted. Also in 1950, the C-37 was introduced. It featured two Chevrolet engines, a raised passenger deck and under floor baggage compartments. It was not successful and only eight were sold. The Clipper styling was somewhat modified when the Starliner was introduced in 1957.

In 1955, the Vistaliner was launched. It was not based on the Clipper design, but was basically a new design with deck-and-a-half seating and under floor baggage compartments. Although diesel engines in Flxible coaches — particularly General Motors 4-71 diesel engines — were offered earlier, a Cummins diesel engine was standard in the Vistaliner. Following the Vistaliner was the Hi-level model in 1959 and the Flxiliner in 1964.

The Flxible Company was very active in the Canadian bus market, and even sold a number of coaches to Mexican bus operators. There were also reports of Flxible coaches being sold to operators in Cuba, Colombia, Argentina, Egypt and elsewhere.

The Australian firm of Ansair Pty., Ltd. became interested in Flxible Clipper coaches for Australia. Ansair purchased a complete Flxible Clipper in 1947, followed by a second vehicle in parts. The first coach was converted to right-hand drive and began service in January 1949. Ansair also obtained a license to build the Clippers in Australia and construction began in 1950. Between December 1950 and June 1960, 131 Ansair Flxible Clippers were built.

Pioneer Tours, a large intercity bus company in Australia and part of the Ansair organization, operated a number of Ansair Flxible Clippers with Leland engines. Later the Ansair Flxibles were re-engined with General Motors 4-71 diesel engines. These Pioneer coaches were in long-distance services linking a number of large Australian cities. Some Ansair Flxible Clippers were exported to New Zealand.

Following the discontinuation of intercity coach manufacturing by the Flxible Company, an agreement was forged between Flxible and the Mexican bus builder Diesel Nacional SA (DINA) to build Flxible intercity coaches, particularly with the Flxiliner style body. These coaches saw service on routes of many Mexican intercity carriers. In recent years, DINA made some design changes resulting in the Dorado and Avante models.

The Flxible Company was involved in some diversification, as it launched manufacture of funeral cars and ambulances at the same time its bus building began in 1924. This continued until 1952, but was revived for five years between 1959 and 1964. In addition, coin-operated lockers were built for use in bus stations and elsewhere.

In 1952, The Flxible Company signed a manufacturing agreement with the Twin Coach Company to produce the Twin Coach city transit buses. Flxible continued to build transit buses after it stopped producing intercity coaches. In the following years, The Flxible Company had two important management changes. A large manufacturing plant was established in Delaware, Ohio, although the Loudonville, Ohio facility was retained and served various functions. Then, in 1995, Flxible ceased all bus building operations.

The Flxible Company had a very rich history, with the Clipper years considered the most significant era for the company. This is when thousands of Clipper-style coaches were built, playing an integral role in the development of the bus industry in America as well as an important role in the transportation history of the United States.

# Early Flxible Coaches

The Flxible Company was established in the early 20th century with motorcycle sidecar manufacturing as its main business. In 1921, mass production of automobiles almost eliminated the demand for sidecars. This shift forced The Flxible Company to seek new opportunities, which it did by launching bus manufacturing in 1924. Pictured here is what has been reported to be the first Flxible bus. It was built on a Studebaker chassis and sold to E.L. Harter, who ran a small bus line in the area.

Prior to 1930 The Flxible Company built sedan-type buses, mainly on stretched Buick chassis. There were doors on the right side of the bus for passengers seated in each row of seats. Twelve passengers could be accommodated. A boot at the rear of the bus could store some of the luggage. Note the plates below each door identifying the bus as a Flxible. It is interesting to observe how styles have changed over the years. *Motor Bus Society*

Flxible built this type of bus in 1930. It had a front entrance and carried 17 passengers. Although it is difficult to recognize the make, it probably was a Buick. Note the louvers above the windows for ventilation. There was also a back door. The back seats were often reserved for smokers, who entered the bus through their own door. Some baggage could be accommodated on the roof. The spare tire was mounted on the fender.

In 1931 and 1932, Buick chassis were used for the buses built by Flxible. Shown here is the Model 17-0, which was unlike previous sedan models. A single door in the front, along with a center aisle, provided the passenger entrance and exit. There was seating for 17 passengers.

# Flxible Airway Coaches

Santa Fe Trail Stages was established in 1933. It was a combination of several bus lines and was owned by the Atchison, Topeka and Santa Fe Railroad. The company was one of the founders of the National Trailways Bus System and operated buses between Chicago and the West Coast. Santa Fe purchased many buses from The Flxible Company, the first being this 16-passenger Airway Model 16-C-75 acquired in 1935.

The Flxible Airway model had a six-cylinder, 60-horsepower Chevrolet engine. Dayton & Southeastern Lines, Inc. added this 16-passenger Airway Model 16-C-75 coach in 1935. The company operated two routes, one between Dayton and Chillicothe and the other between Springfield and Chillicothe, Ohio.

The Flxible Airway coach was introduced in 1934. The body was mounted on a 1934 Chevrolet truck chassis equipped with special springs and other features making the chassis suitable for bus use. The first models seated 16 passengers. L.A. Nance Bus Lines of Ada, Oklahoma, which operated considerable service across the southern portion of that state, was one of the first bus companies to buy the new Airway Model 16-C-75 in early 1935.

Approximately 130 16-passenger Flxible Airway coaches were built between 1934 and 1938. Bee Line Coaches of Waco, Texas, purchased this Flxible Airway Coach Model 16-C-74 in 1934. Bee Line Coaches operated for a short time on a route between Waco and Palestine, Texas.

The 19-passenger Flxible Airway coach was introduced in 1936, with Southwestern Stages of Mankato, Minnesota, receiving the first one. It was identical to the 16-passenger Model 19-C-76 but was 341 inches long instead of 305 inches. Southwestern Stages began in 1933 when W.L. Jaax became the owner of Blackhawk Transportation Company, operating between Mankato and Sioux Falls, South Dakota. For several years, Jaax was the Flxible salesman in the area.

Interstate Buses Corporation of West Springfield, Massachusetts, purchased three Model 19-C-76 Flxible Airway coaches in 1936. These buses had a number of special refinements such as chrome rear bumpers, spotlight, reflectors and rear-view mirrors. Interstate operated several schedules daily between Providence, Rhode Island and Albany, New York. Bonanza Bus Lines purchased the company in 1957. It was later acquired by Coach USA in 1998 and by Peter Pan Bus Lines in 2002.

The Flxible Company built many special coaches for businesses and private individuals over the years. This 1936 Flxible Airway model was one of the early examples. It was built on a Chevrolet chassis with a 16-passenger body and sold to Jim Handy Motion Picture Service, Detroit, Michigan, for use in promotions.

The Flxible Airway coaches were mounted on newer Chevrolet chassis in 1938. Note the front fenders and grille are different than previous Airway models. Columbus Marysville Bus Co., Inc. of Columbus, Ohio, which owned this 1938 Flxible Airway, served Columbus, Marysville, Kenton and Bellefonte. Bus companies in the 1930s carried the names of towns served, and some cities served by connection, on the belt below the windows. This Model 19-C-78 Airway was delivered in 1938.

Crescent Stages of Anniston, Alabama purchased three Flxible Airway coaches in 1936 and added this newer Airway Model 19-C-78 in 1938. Most Flxible Airway coaches had a space in the rear roof for baggage, which is seen on this model. Crescent Stages was a good Flxible customer for many years and became a member of the National Trailways Bus System in 1940.

# Original Flxible Clipper Coaches

In 1937, the new line of Flxible coaches received the name Clipper. It had a body similar to the Airway model, but the engine, still located in the front, was under the body next to the driver. Southeastern Stages of Atlanta, Georgia purchased one of the first Clippers, Model 20-CL-77. Southeastern had several Flxible Airway coaches prior to adding this new Clipper. The bus company was formed in 1933 and operated a number of routes in Georgia. It continues to operate today, celebrating 70 years in 2002.

Turner Transportation Company of Shawnee, Oklahoma had two Flxible Clipper Model 20-CL-77 coaches for its two north-south routes in Oklahoma. The new Clipper coaches feature the same engine and chassis as in the Airway models. Bendix-Westinghouse air brakes were standard on the new Clippers. Turner also had two Airway coaches and another Clipper.

Flxible Model 20-CL-77 coaches were very popular among bus companies in Texas. Ten Texas bus companies had Flxible coaches in their fleets by the end of 1938. Airline Motor Coaches of Nacogdoches was one of these 10, and one of the first to buy a Model 20-CL-77 Clipper, taking delivery in early 1937.

Panhandle Stages of Amarillo, Texas began in 1934 with several routes east and northeast from Amarillo. The route between Amarillo and Oklahoma City was an important link east and west, which led Panhandle to become a member of the National Trailways Bus System in 1936. This Flxible Clipper Model 20-CL-77 was purchased in September 1937. The Model 20-CL-77 Flxible had an overall length of 334-1/2 inches and was 88 inches wide.

Prost Bus Line of Perryville, Missouri purchased this Flxible Clipper Model 20-CL-77 in November 1937. It was used by the company on its route through various small towns between St. Louis and Cape Girardeau, Missouri. Prost also purchased a 25-passenger Flxible Clipper coach in 1941.

Flxible coaches were sold to bus operators throughout the country. River Auto Stages of Sacramento, California purchased this Clipper Model 20-CL-77 in 1937. River Auto Stages had a number of routes between Sacramento and Maryville, Stockton, Rio Vista and other nearby cities. Several other companies in California also purchased Flxible coaches.

The 20-passenger Flxible Clipper coaches featured high-back seats, two on the left side and one seat on the right, trimmed with a combination of leather and mohair. In September 1937, Salt Lake & Tooele Stages purchased this Clipper Model 20-CL-77 coach. The company, launched in 1920 and folded in 1954, operated between Salt Lake City and Grantsville, Utah.

There were 115 Flxible Clipper Model 20-CL coaches built in 1937 and 1938. Many small bus lines acquired this model. Eastern Shore Stages of Selbyville, Delaware was one of these small bus lines, which purchased this Model 20-CL-77 Clipper in September 1937. Eastern Shore Stages operated one daily round trip between Wilmington and Selbyville, a distance of 82 miles.

25

Most orders for Flxible Airway and Clipper coaches were single-unit orders. One order for three Clipper 20-CL-78 coaches came from St. Andrews Bay Transportation Company, commonly called The Bay Line, of Dothan, Alabama. The Bay Line served Dothan as well as the Florida cities of Pensacola and Panama City. These three Clippers were purchased in late 1938. The company was sold to Georgia Stages, a Trailways member, in 1944.

26

# Rear-engined Flxible Clipper Coaches

The Flxible Company introduced its first rear-engined coach, the Clipper Model 25-CR-78, in June 1938. It was new in many respects, including an all-metal chassisless design. Peninsula Transit Corporation of Norfolk, Virginia ordered two of these new units soon after they became available. Two more were added to Peninsula's fleet in 1938. Peninsula had an important route between Norfolk and Richmond, which was later taken over by Greyhound.

Jack Rabbit Lines of Sioux Falls, South Dakota added this Flxible Clipper Model 25-CR-78 in the summer of 1938. Jack Rabbit was a pioneer bus company in South Dakota, originating in 1924. The company served a number of South Dakota communities, as well as some in North Dakota and Minnesota. The new Flxible 25-CR-78 had a Chevrolet 78-horsepower engine mounted in-line at the rear of the coach. Jack Rabbit purchased two more 25-CR-78 Clippers and added three larger models at a later date.

Indianapolis Vincennes Coach Co., Inc. of Vincennes, Indiana purchased four Flxible Clipper Model 25-CR-78 coaches in 1938. The company had a route between Vincennes and Indianapolis, but discontinued route service in recent years. Now known as I-V Coaches, the company continues to operate charters today. The Flxible Clipper Model 25-CR had specially designed seats by the Art Rattan Works with Firestone foam rubber seat cushions and backs.

Arrow Coach Lines of Brownwood, Texas added these two Flxible Clipper 25-CR-78 coaches to its fleet in 1938. A good view of the front of the 25-CR model is shown in this picture. Seven chrome strips appeared below the Flxible name, and there were two fog lights just above the bumper. Arrow Coach Lines had two routes from Brownwood, one serving Austin and the other serving Waco. The company later moved to Killeen, Texas and continues operating in Texas today as a Trailways company.

Gulf Transport Co., Mobile, Alabama, purchased this Flxible Clipper 25-CR-78 model in 1938. Gulf's first bus fleet consisted of four Flxible Airway coaches. Gulf Transport, officially established on August 12, 1936, was a subsidiary of the Gulf, Mobile & Ohio Railroad. Its main route was from Mobile to St. Louis, but it also had other secondary routes. It was sold in 1985.

Indiana Motor Bus Company, Plymouth, Indiana, purchased two or more Flxible coaches each year from 1938 until World War II temporarily stopped bus production. Two of the Flxible Clipper Model 25-CR-79 coaches are shown here in December 1939. One of the features of the rear-engined Clippers was an enclosed, 150-cubic-foot baggage area at the rear. Indiana Motor Bus Company was a pioneer northern Indiana bus company. It sold to United Limo in 1984.

In late 1939, Buick 141-horsepower engines mounted in-line in the rear were made available in Flxible Clipper coaches. The Buick engine was a straight-eight design and provided more power and stamina. It also had a very distinct sound. Moyers Stages of Fresno, California, which operated several routes in the area, acquired three of the first Flxible Clipper 25-BR-149 Buick-powered coaches.

The Flxible Company had its first Canadian order in 1938, with two Flxible Clippers sold to Gray Coach Lines of Toronto, Ontario. The next year, seven units were sold to Acadian Lines in Halifax, Nova Scotia. In 1939, Wagner Tours, Ltd. of Yarmouth, Nova Scotia purchased three Model 25-BR-149 Buick-engined Clipper coaches. Pictured here are the three shown from the left side. Note the emergency door and the louvered rear window where the baggage compartment was located. Wagner operated service between Yarmouth and Halifax and a second route between Yarmouth and Bridgewater.

Several Indiana bus companies operated Flxible coaches. One was Indianapolis and Southeastern Lines of Indianapolis, which began in 1920 and continues in business today as a charter company. The line originally operated between Indianapolis and Cincinnati and later expanded to Chicago and Knoxville, Tennessee. This Flxible Clipper Model 25-CR-79 was photographed in 1939 before the company joined the National Trailways Bus System later that year.

This Flxible Clipper 21-CR was an unusual model, although it looked the same as the Model 25-CR. It had seats for 21 passengers instead of 25 because the rear bulkhead between the passenger and baggage compartments was moved forward to provide more baggage and package space. Only 13 Model 21-CR Clippers were built. Red Ball Lines of Huron, South Dakota took delivery of this Clipper Model 21-CR-70 in 1940. Red Ball, one of several small bus lines in the Dakotas, operated service linking the cities of Mitchell and Pierre.

Vermont Transit Co. of Burlington, Vermont had five Flxible Clipper 25-BR-140 coaches, the last three being delivered in mid-1940. These were among the last of the square-window-style Clippers. Vermont purchased several more Flxible Clippers the following year. Vermont Transit was a pioneer bus company in New England, beginning in 1929. It was acquired by Greyhound in 1975 but continues to operate in New England today under the Vermont Transit name and the traditional green-and-white color scheme.

# Rear-engined 29-passenger Flxible Clipper Coaches

The Flxible Company introduced air-conditioning on its coaches in June 1939. Pictured here is a Flxible Clipper 29-BR-149 coach delivered to Short Way Lines of Toledo, Ohio in September 1939. The generator and compressor for the coach air-conditioning system was belt-driven from a four-cylinder, 12-horsepower Waukesha engine mounted in the rear baggage compartment. Note the louvers at the rear and the service door. The Frigidaire Corporation made the air-conditioning units especially for Flxible. Short Way Lines had an extensive route system from Toledo north into Michigan.

The Flxible Company introduced the 29-passenger Clipper coach in early 1939. It was similar to the 25-passenger Flxible Clipper but had a 218-inch wheelbase instead of the 182-inch wheelbase of the 25-passenger model. The longer Clipper, which had a Buick engine, was designated the Model 29-BR. Carolina Coach Company was one of the first purchasers, ordering five 29-BR-149 units in 1939. This was soon after Carolina joined the National Trailways Bus System and buses were painted in Trailways' ivory and crimson colors. Carolina continues to operate today and has been owned by Greyhound since 1997.

Bowen Motor Coaches of Fort Worth, Texas was an important Flxible customer, purchasing 13 air-conditioned Flxible Clipper 29-BR-149 coaches (one pictured here) in 1939. The previous year, four Model 25-CR Clippers were acquired. Bowen was one of the first carriers to buy the 29-passenger Flxible Clipper. Bowen Motor Coaches was created in 1936 from an amalgamation of several Texas bus companies in 1936, and became a National Trailways Bus System member in 1943. Twenty companies purchased a total of 69 units in the 29-BR's first year of production.

Utica-Rome Bus Co., Inc. of Utica, New York added this Model 29-BR-140 Flxible Clipper coach in 1940. It was the first large bus for the company, which began in 1933. The Utica-Rome Bus Co. route was approximately 15 miles. A side destination sign was ordered for this bus at an additional cost of $50. The Buick-engined, 29-passenger Clippers were popular with both large and small bus companies.

# New Design Flxible Clipper Prototype

In April 1940, a new Flxible Clipper known as the Model 29-BR-140B (prototype pictured here) was introduced. Instead of side windows that were rectangular, the new windows were the sliding type and had a slanted shape. Fluted stainless-steel paneling also decorated each side of the coach. Production models had a changed front end featuring three chrome bumpers identifying the new model. Both 29-passenger and 25-passenger models were produced.

# New Design 25-passenger Flxible Clipper Coaches

The new 1940 model Flxible Clipper coaches were available in both 29-passenger and 25-passenger sizes. The smaller Clippers offered either a Chevrolet or Buick engine. Northern Transportation Company of Bemidji, Minnesota purchased this Flxible Clipper 25-BR-140 model with a Buick eight-cylinder engine in 1940. Northern, established in 1936, operated two routes in northern Minnesota: between Bemidji and International Falls, and Virginia and International Falls with a branch to Baudette.

Las Vegas-Tonopah-Reno Stage Line of Las Vegas, Nevada began in 1930 with a sedan bus running between Las Vegas and Tonopah, with Reno eventually added to the route. Two of the company's first buses were these two 1940 Clipper Model 21-CR-70B coaches, with room for 17 passengers and extra rear baggage space. Window guards were placed inside the last windows on each side to protect the baggage. These two Clippers were air-conditioned and welcomed by Nevada passengers.

The 1940 Flxible Clipper coaches were made available with either a Chevrolet or Buick engine. Orange Belt Stages of Visalia, California chose the Chevrolet engine for this and one other Clipper Model 25-CR-70 it purchased in 1940. Orange Belt had its beginnings in 1918 but didn't adapt that name until 1934, and through 1948 had purchased more than 50 Flxible coaches. Orange Belt Stages continues to operate route services and charters today.

45

Triangle Transportation Company began in 1921 in Crookston, Minnesota, and served a large section of northwestern Minnesota. In 1940, Triangle purchased this Buick-powered Flxible Clipper Model 25-CR-140B. This was the seventh Flxible in the Triangle fleet, including three Airway models. The area served by Triangle had cold winters that provided a real test for Flxible coaches. A storm sash was made available for each of the passenger windows. Triangle, now based in East Grand Forks, Minnesota, continues today as a charter carrier.

The new 25-passenger Flxible Clipper was very popular, being operated by many companies including Consolidated Bus Lines of Bluefield, West Virginia. Consolidated, as its name suggests, was formed from a combination of many small companies in southern West Virginia. The Flxible Clipper 25-BR-41 was one of the four acquired by Consolidated in 1941. Consolidated purchased nine more Flxible coaches through 1946. The company joined the National Trailways Bus System in 1954, and then sold to Virginia Trailways in 1955.

Although the Bowling Green-Hopkinsville Bus Company served the two cities in its name, the Russellville, Kentucky-based bus company served several other cities in Kentucky and Tennessee, including Owensboro, Kentucky and Clarksville, Tennessee. The Bowling Green-Hopkinsville Bus Company purchased this Flxible Clipper 25-CR-41 in June 1941. Previously the company had one or more of each of the Flxible models, beginning with the Airway model purchased in 1935.

Gator Motor Lines of St. Augustine, Florida took delivery of this Flxible Clipper 25-CR-41 in August 1941. Although the Florida Motor Lines name appears above the windows, Gator Motor Lines, the owner of the bus, is named above the door. Gator Motor Lines operated a short route between Gainesville and St. Augustine. The Flxible Clipper 25-CR-41 model weighed 10,400 pounds.

48

New Mexico Transportation Co., Roswell, New Mexico, purchased these two Flxible Clipper Model 25-BR-41 coaches in 1941. New Mexico Transportation, which served most of New Mexico as well as Amarillo and El Paso, Texas, had been a Flxible customer since 1936 when four Model 16-C-76 coaches were purchased. The company then added one or more Flxible coaches each year until production ceased for World War II.

A frequent buyer of Flxible coaches was Queen City Coach Co. of Charlotte, North Carolina. Pictured here is a Queen City Flxible Clipper Model 25-CR-41 acquired in June 1941, one of 82 Flxible coaches purchased by Queen City between 1935 and 1942. The first order in 1935 was for three Flxible Airway Models 16-C-75. The company served most cities in southern and western North Carolina, and joined the National Trailways Bus System in 1940.

# New Design 29-passenger Flxible Clipper Coaches

The first Flxible 29-passenger Clipper models introduced in 1940 had paired, slanted side windows, and later all windows — still slanting — were the same. The difference is illustrated with photos of the two coaches from the Little Rock-based Arkansas Motor Coaches fleet. The Flxible Model 29-BR-140B (top) shows the paired windows, and the Flxible Model 29-BR-41 (bottom) depicts the updated windows. The company was a good customer for Flxible coaches, buying its first units in 1937. Arkansas Motor Coaches had a main route between Memphis, Tennessee and Texarkana, Texas.

These four Flxible Clipper Model 29-BR-140B coaches had an unusual feature not recognized on other deliveries. These coaches had only two instead of three bumpers, and a front end like the previous one-bumper Flxibles. Paired, slanting windows were also featured. Syracuse and Oswego Motor Lines of Syracuse, New York operated frequent service between the cities in its name. It began in 1931 and continues to operate motor coaches today under the name S&O Tours.

In the early 1940s The Flxible Company reported that most orders for Flxible Clipper coaches were repeat orders. Many were multiple orders as well. Tri-State Transit Co. of Shreveport, Louisiana was a frequent Flxible customer. When these four Flxible Clipper 29-BR-41 coaches were delivered in October 1941, Tri-State Transit had purchased a total of 67 Flxible coaches. The first two were Flxible Airway Model 16-C-76 coaches, delivered to Tri-State in 1936. Texas, Louisiana and Mississippi were the primary states served by Tri-State.

Missouri, Kansas & Oklahoma Coach Lines of Tulsa, Oklahoma purchased three Flxible Clipper 29-BR-41 coaches in 1941, two of which are pictured here. These coaches were air-conditioned and featured Buick 165-horsepower straight-eight engines. The new engines were more powerful than the Buick engines introduced in 1940. Missouri, Kansas and Oklahoma Coach Lines had a main route between St. Louis and Oklahoma City, as well as many short routes in Arkansas and Oklahoma. The company became a Trailways member in 1938.

These three Flxible Clipper Model 29-BR-41 coaches were added to Oklahoma Transportation Company's fleet in June 1941. They, like many Flxible coaches purchased at that time, were air-conditioned. Oklahoma City-based Oklahoma Transportation Company operated service throughout the central part of the state, with a main route between Fort Smith, Arkansas, and Wichita Falls, Texas. The company began in 1929 and was acquired by Jefferson Lines in 1982.

In the 1940s, The Flxible Company emphasized standardized parts and materials and ease of maintenance in its advertising, a strategy that attracted many customers. Mount Hood Stages of Bend, Oregon was one, acquiring this Flxible Clipper Model 29-BR-41 coach in June 1941. It was the company's third Flxible, with 41 more to follow through 1946. Mount Hood Stages began in 1931 and joined the National Trailways Bus System in 1943, eventually becoming Pacific Trailways. The company ceased business in 1982.

This Flxible Clipper Model 29-BR-41 joined the fleet of Reading (Pennsylvania) Transportation Company in 1941. It had three bars at each side window, fitted for $3 per window. Reading Transportation, a Reading Railroad subsidiary launched in 1926, operated passenger service on many routes originally served by rail in eastern Pennsylvania and New Jersey. The company joined the National Trailways Bus System in 1955 and was sold to numerous Pennsylvania bus companies in 1964.

Western New York Motor Lines of Batavia, New York purchased this and one other Flxible Clipper Model 29-BR-41 in 1941. Western New York Motor Lines used the trade name Blue Bus Lines. The coaches owned by the company were painted blue and white and also sported an unusual checkerboard design in front. The route was between Buffalo and Rochester, New York. The company became a Trailways member after World War II, adopting the name Empire Trailways. It is today known as New York Trailways.

Gray Coach Line of Toronto, Ontario purchased 48 Flxible coaches between 1938 and 1948. The first two had Chevrolet engines and the remainder were Buick-powered. Pictured are four of the 10 Flxible Clipper Model 29-BR-41 coaches added in 1941. Gray Coach, which began in 1927, was a subsidiary of the Toronto Transportation Commission and served many cities in southern Ontario. The Gray Coach system was sold to Greyhound Canada in December 1992.

The Orange Line, with its colorful orange and black livery, purchased these Flxible Clipper Model 29-BR-42 coaches in March 1942. The Orange Line was the trade name for the bus line between Green Bay, Wisconsin, and Dubuque, Iowa, operated by the Madison, Wisconsin-based Wisconsin Power and Light Company. Bus service by the power company was sold to Northland Greyhound Lines in 1945. These Flxibles were the only ones purchased by The Orange Line.

Capitol Bus Company of Harrisburg, Pennsylvania purchased its first Flxible coach, a Clipper Model 20-CL-77 in 1937, a year after the company began, and it was the third coach in the fleet. The Flxible Clipper Model 29-BR-41 pictured here was purchased in 1941. It was the seventh Flxible acquired by Capitol. Thirteen more were purchased later. Capitol Bus Company joined the National Trailways Bus System in 1948 and continues as a Trailways member today, serving Pennsylvania as well as Baltimore, Maryland and Washington, D.C.

Great Southern Coaches, Inc. of Jonesboro, Arkansas purchased two Flxible Clipper Model 29-BR-42 coaches in September 1942. They were the first Clippers in the Great Southern fleet. The company had its beginnings in 1938 and had a main route between St. Louis, Missouri, and Memphis, Tennessee. At Jonesboro, there was a branch line to Little Rock, Arkansas. Great Southern Coaches continues today as a charter-only operator.

# Small 18-passenger Flxible Coaches

Two Flxible Model 18-CF-41 coaches were sold in June 1941 to Dakota Bus Lines of Fargo, North Dakota, with another added later that year. Dakota Bus Lines, a small bus line serving North and South Dakota, purchased its first Flxible — an Airway model — in 1937. Only 56 of these small Flxibles were built. They had a wheelbase of 187-5/8 inches and an overall length of 27 feet.

Logan Williamson Bus Company of Logan, West Virginia operated several routes in southern West Virginia. It purchased this Flxible Model 18-CF-41 — the eighteenth Flxible coach added to the Logan Williamson fleet — in 1941. The company had its start in 1926 and was sold 20 years later to Consolidated Bus Lines. The Flxible 18-CF-41 had a 235-cubic-inch Chevrolet six-cylinder heavy-duty engine mounted at the front next to the driver.

In May 1941 The Flxible Company introduced a new 18-passenger coach designated the 18-CF-41, featuring a die-formed steel skeleton. The all-metal body and chassis were welded together as a single unit. One of the first bus companies to order the new coach was United Motor Ways of Grand Island, Nebraska. Pictured here is the company's Flxible 18-CF-41 coach. United Motor Ways operated several central Nebraska routes, each with a different owner.

Shown here is the front interior of the Flxible Model 18-CF-41 coach. The Chevrolet engine was located to the right of the driver. Note the long reach of the gearshift and the door opener. The entrance/exit door was located behind the right front wheel, and two fans were located at the windshield. These fans cost $8 each and were used for defrosting the windshield. The spotlight was $20 extra.

# Flxible Clipper Coaches During World War II

Some wartime restrictions were imposed on the bus industry in 1942, such as the simplified paint scheme and lack of stainless-steel side paneling on this Flxible Clipper Model 29-BR-42 acquired at that time by Portland-based Oregon Motor Stages. Most coaches at that time had a similar appearance. Oregon Motor Stages was a large user of Flxible coaches, purchasing 51 units between 1939 and 1944. The company began in 1931 and had a number of routes in northwestern Oregon and along the Pacific Coast. Pacific Greyhound Lines acquired Oregon Motor Stages in 1954.

Zephyr Lines of Minneapolis, Minnesota purchased this Flxible Clipper Model 29-BR-44 coach in 1944. It was one of the first Flxible coaches produced following almost two years during which the Flxible plant devoted all its attention to producing products for the war effort. Zephyr Lines operated only Flxible Clipper coaches from its start. It operated between Minneapolis and Ashland, Wisconsin. Note that this Flxible had rub rails at the lower part of the body. These were added extras, priced at $45 for each side.

This Flxible Clipper Model 29-BR-45 was one of three coaches purchased by Mac Kenzie Bus Line, Ltd. of Bridgewater, Nova Scotia in 1945. One similar model was acquired the year before. Mac Kenzie began in 1933 with a Halifax-Bridgewater route, with the line eventually extending to Yarmouth, Nova Scotia. The company discontinued service in October 1998.

Greyhound companies were among the best customers for Flxible coaches. Dixie Greyhound Lines of Memphis, Tennessee purchased the two Flxible Clipper 29-BR-45 coaches pictured here in 1945. Dixie Greyhound had purchased a total of 33 Flxible coaches. Teche Greyhound Lines of New Orleans, Louisiana had acquired the greatest number of Flxibles by Greyhound companies, a total of 53 in all, beginning with two Airway models purchased in 1936. There were 278 Flxible coaches operated by all companies carrying the Greyhound name.

# Post-war Flxible Clipper Coaches

Canada Coach Lines, Ltd. of Hamilton, Ontario took delivery of 15 Flxible Clipper Model 29B-147 coaches in 1947. Two similar coaches had joined the fleet the year before and 22 more were added later, making a total of 39 new Flxible coaches operated by Canada Coach Lines. The company, a subsidiary of the Hamilton Street Railway Company, operated a number of routes in southern Ontario. Trentway-Wagar, Inc. purchased Canada Coach Lines in 1993.

Acadian Lines of Halifax, Nova Scotia was a very good Flxible customer. The company, which served much of Nova Scotia, got its start in 1936. The first Flxible coaches were acquired in 1939. A total of 65 Flxible Clippers and Visicoaches were purchased by Acadian, including this Flxible Clipper Model 29B-147 added in 1949. Many Canadian bus companies used Flxible coaches.

Maine Central Transportation Company of Portland, Maine purchased these three 33-passenger Flxible Clipper Model 33B-149 Suburban Express coaches in 1949. Instead of having normal space for baggage in the rear, seating was extended rearward to accommodate an extra row of seats. Only two other Flxible coaches, both 25-passenger models, were in the Maine Central fleet. The company, a subsidiary of the Maine Central Railroad, began in 1925 and ceased operations in 1956.

This Flxible Clipper Model 23B-247 was one of the coaches in a three-bus order for the Provincial Transport Company, Montreal, Quebec. These coaches seated 23 passengers in a two-and-one configuration instead of the normal 29 passengers, and were used on Laurentian Mountain Tours. Provincial had previously purchased 25 Flxible Clippers. Following this order, 20 more Clippers joined the Provincial fleet. Provincial Transport began in 1928 and served many intercity and suburban routes in Quebec. Provincial, which later adopted the name Voyageur, sold its routes to Orleans Express and several other companies in 1990. *Paul A. Leger collection*

Colonial Coach Lines, Ltd. of Ottawa, Ontario had 30 Flxible coaches in its fleet; however, they were not acquired directly. Several 29-passenger Flxibles built in 1947 and 1948 were originally in the Montreal-based Provincial Transport Co. fleet. The Flxible Clipper Model 29B-147 pictured here was a 1947 model. Although Colonial Coach Lines kept its identity, it was a part of the Provincial Transport network. The name Voyageur was later adopted for both companies.

73

Rock Island Motor Transit Company of St. Paul, Minnesota, a subsidiary of the Chicago Rock Island and Pacific Railroad, purchased this Flxible Clipper Model 29B-149-AC in November 1949. The Flxible Company sold a number of coaches to railroad-owned bus services. The Rock Island Motor Transit coach was used to transport passengers between Rochester and Owatonna, Minnesota where the railroad had a line. Rock Island Motor Transit later purchased a Flxible Visicoach.

In September 1949 the Worcester (Massachusetts) Street Railway Company purchased eight Flxible Suburban Express Model 37B7-49 coaches for use on several suburban routes operated by the company. The Suburban Express coach had 37 high-back non-reclining seats. An air-operated folding door was specified for the coaches in this order. The Worcester coach pictured here was the 3,000th post-war Flxible coach. Officials of Flxible and Worcester are pictured with this coach. *Motor Bus Society*

Taken in front of the Flxible plant in Loudonville, Ohio, this photo from a March 1949 advertisement featured the four types of coaches that were being sold at the time. Shown left to right are the Model 21-C1 Airporter, the Model 25-C1 powered by Chevrolet, the Model 29-B1 Clipper powered by Buick, and the Model 37-C1 Clipper with two Chevrolet engines. The advertisement emphasized that Flxible built coaches for various types of services in several sizes.

# C-1 Dual-engined Flxible Clipper Coach

The Flxible Clipper Model C-1 coach, first introduced in June 1950, was a 37-passenger Hi-level model. The most interesting feature of this coach was that it had two Chevrolet Loadmaster 471-cubic-inch, six-cylinder gasoline engines. A transfer case connected the engines to the transmission and other components. Only eight Model C-1 units were built. Iowa State Teachers College of Cedar Falls, Iowa purchased the one pictured here.

# Flxible Post Office Coaches

In 1950, The Flxible Company joined with a number of bus manufacturers to build special buses — referred to as Highway Post Office vehicles — for the United States Post Office Department. Flxible is reported to have built 11 of these vehicles in 1950. They were sold to individual bus companies that had contracted to operate them for the Post Office Department. They were similar to the Flxible Clipper buses, but had a higher roof and a very different interior because employees would sort mail as the vehicles traveled between various communities on specified routes.

# Flxible Sightseeing Coaches

In 1947 this Model 25B1-47 Flxible sightseeing coach went into service in the Denver, Colorado area. It was operated by Gray Line of Denver, which was part of an organization operating several bus and taxi services. This particular sightseeing coach had large side windows and a glass top. A rear emergency door was required, making it necessary to have two windows instead of one large window to match the others.

The Flxible Company built several styles of sightseeing coaches. This Model 33B6-48 unit and two others were acquired by Salt Lake Transportation Co. of Salt Lake City, Utah in 1948. These coaches, like many other Flxible sightseeing coaches, had large rectangular side windows. The Salt Lake coaches had sliding windows and accommodated 33 passengers. Salt Lake Transportation Co. was a member of the Gray Line Sight-Seeing Association at the time.

Yosemite Transportation System in Yosemite National Park, California purchased its first six Flxible coaches in March and April 1949. The Flxible Visicoach shown here was added in June 1952. At that time there were no ventilation intakes at the front, and the stainless-steel paneling extended only to the middle of the door. A Fageol Model FTC-180, 404-cubic-inch gasoline engine, powered this coach. In addition to using the Visicoaches for tours within the park, the company had routes that served Merced and Fresno, California from Yosemite.

The Flxible Company built a number of coaches with glass tops for sightseeing purposes. Rocky Mountain Tours Co., Ltd. of Banff, Alberta purchased two of these Model 218-B6-33SS Flxible Visicoach sightseeing coaches in May 1951. The company was sold to Brewster Transport in September 1957 and renamed Brewster-Rocky Mountain Gray Line.

# Flxible Visicoaches

One of the first Flxible Visicoach models built was this Model 218-B1-50-29-IC, delivered to the Rochester Penfield Bus Co., Inc. of Rochester, New York in August 1950. Buyers of the first Visicoach models had a choice of a Buick FB Fireball 152-horsepower gasoline engine or a six-cylinder, four-cycle Hercules Model DWXLD 142-horsepower diesel engine. The Rochester-Penfield Bus had a Buick engine. The company used the trade name Valley Bus Lines, served several suburban communities, and had a route connecting Rochester with Elmira, New York.

Zanesville, Ohio-based Zane Transit Lines acquired two Model 218-B1-50-29-1C Flxible Visicoaches in September 1950. They were the eleventh and twelfth Flxible coaches added to the Zane Transit fleet, with the first acquired in 1936. These Visicoaches continued to feature ventilation louvers at the front. The new Visicoaches had Buick FB straight-eight, 144-horsepower gasoline engines. Zane Transit Lines had a single route in Ohio running between Canton and Marietta via Zanesville.

Interstate Transportation Company of Minot, North Dakota was one of the state's pioneer bus companies, with a main route between Bismarck and Minot as well as several other routes. The company operated a number of Flxible coaches, beginning with two Model 18-CF-41 Flxible coaches in 1941. The Flxible Visicoach Model 218B-150 pictured here was purchased in 1950. It had a Buick 152-horsepower engine and a Spicer five-speed synchromesh transmission.

The Flxible Visicoach was introduced in 1950. Its main feature was the longer and higher side windows, and there also were air vents on each side of the coach at the front. Okanogan Valley Bus Lines of Spokane, Washington purchased this Model 218-B1-51-29-IC Visicoach in April 1951. The company began with a short route in the Okanogan Valley in Washington and later operated between Oroville, Washington and Spokane.

This Flxible Visicoach, Model 218-GM1-50, was purchased in 1950 by the Berea Bus Line Co. of Berea, Ohio. At the time Berea Bus Line Company operated city bus service in the Cleveland suburb of Berea, and also had a route into Cleveland and other neighboring communities. Note that this Flxible had no stainless steel trim below the windows but there was a decorative strip at the skirt. Berea Bus Lines originated in the 1920s and was sold to the Cleveland Transit System in January 1968 for $400,000, which included the bus line, 23 transit buses and 27 school buses.

Badger Coaches of Madison, Wisconsin purchased this Flxible Visicoach Model 218-GM1-53-29-IC in June 1953. Originally it had a Buick engine, but later a General Motors 4-71 diesel engine was installed. After five years Badger Coaches, which operated a Madison-Milwaukee service, transferred the coach to an associated company, Badger Bus Company, that operated a Madison-Freeport, Illinois route. Badger Bus Company's route service began in 1920 and ended in 1982. Badger Coaches continues today, and Badger Bus Company only operates school bus service.

87

Boise-Winnemucca Stage Lines of Boise, Idaho purchased this Flxible Visicoach, Model 218FA-1-53-29IC, in June 1953. It was one of several Flxible Visicoaches operated by the company, which had a three-state route between Boise and Winnemucca, Nevada. The Flxible Company built approximately 1,000 Visicoaches between 1950 and 1957. Boise-Winnemucca Stages, founded in 1943, continues to operate today as a charter-only company.

In April 1953 The Flxible Company delivered 16 Visicoaches to Northland Greyhound Lines in Minneapolis, Minnesota. These coaches were unique in two respects: They had retractable wheels at the rear, and also had General Motors diesel engines. The model designation was 218GM1-53-29IC. The extra wheels were needed because, at the time, Minnesota had strict weight restrictions imposed on a number of highways that Northland Greyhound traveled in spring months when the ground thawed. The extra wheels supposedly gave the coach better weight distribution. After the spring thaw the wheels could be retracted.

The White Motor Company of Canada marketed Flxible coaches in Canada for a time following World War II, and many of the coaches had White gasoline engines. Three Flxible Visicoaches were sold through White Motor Company of Canada to Autobus Drummondville Ltd., Drummondville, Quebec. This 1951 Drummondville Visicoach, Model 33B1-51, had 33 seats and a White engine.

One of the last Flxible Visicoaches built was this Model 218GM1-56-29-ICAC delivered to River Trails Transit Lines of Dubuque, Iowa in May 1956. It was air-conditioned and had new-style turn signal lights. River Trails had a number of intercity routes in eastern Iowa. It was established in 1939 by Joseph Wenzel, and sold to James Hillard of Galena, Illinois in 1974. It continues today as a tour and charter company using the Tri-State Travel name.

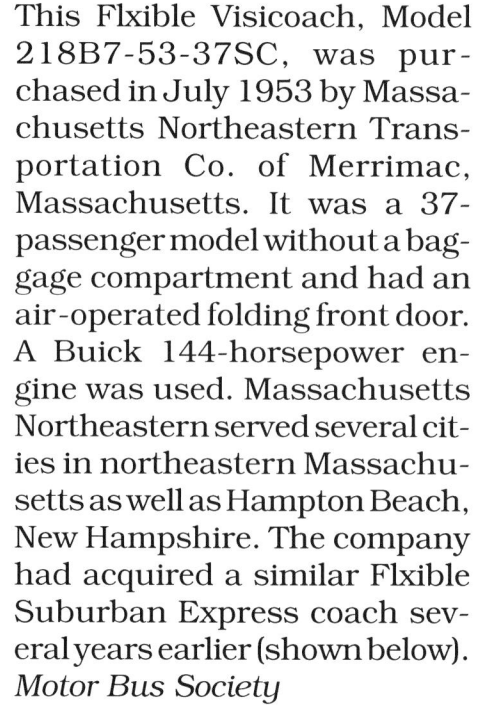

This Flxible Visicoach, Model 218B7-53-37SC, was purchased in July 1953 by Massachusetts Northeastern Transportation Co. of Merrimac, Massachusetts. It was a 37-passenger model without a baggage compartment and had an air-operated folding front door. A Buick 144-horsepower engine was used. Massachusetts Northeastern served several cities in northeastern Massachusetts as well as Hampton Beach, New Hampshire. The company had acquired a similar Flxible Suburban Express coach several years earlier (shown below). *Motor Bus Society*

# Flxible Airporter Coaches

The Flxible Company began building coaches especially for airport service in 1946. They featured two-and-one seating and often had enlarged baggage space in the rear. In May 1954 two small Flxible Airporter Model 182-B1 coaches, seating 18 passengers, were delivered to the Dixie Traction Company of Covington, Kentucky. They were used for airport transfer service between the Cincinnati/Northern Kentucky Airport in Kentucky and downtown Cincinnati, across the Ohio River. Dixie Traction Company was a part of the Cincinnati Newport & Covington Railway Company.

Continental Air Transport Company, Chicago, Illinois was formed as a division of the Parmelee System in 1943 for transferring airline passengers to and from Chicago's airport and the downtown. Continental began using Flxible coaches at the time and became a very good customer. Between 1951 and 1956, 24 Flxible Visicoaches were added. One is pictured here in downtown Chicago. Later, Flxible Starliners were used and there were also the larger Flxiliners in the fleet. The Parmelee System began in 1853 for transporting passengers between Chicago's several railway stations. The 150th anniversary of the company was celebrated in 2003.

# Flxible Starliner Coach-

The Flxible Company introduced its last Clipper-style coach, the Starliner, in 1957. It was very similar to the Visicoach, but did not have the stainless steel trimming on the sides. However, it had a windowed hump midway at the roof. Airport Transfer Co., Ltd. of Halifax, Nova Scotia, acquired two of these Starliner coaches in September 1957. They were Models 218WAF57-29IC, originally with White engines. The White engines were replaced with General Motors 4-71 diesel engines in later model years. Airport Transfer Co., Ltd. provided service between Halifax and the city's airport.

When sales of coaches to various bus operating companies slowed after World War II, The Flxible Company began marketing its coaches to colleges and universities for their transportation needs. In May 1957 this Flxible Starliner Model 218-GM1-57-29XC was delivered to the Illinois State Normal University, Normal, Illinois. It had a General Motors 4-71 diesel engine. Schools such as the University of Michigan, Xavier University, the University of Miami and many others purchased Flxible coaches.

The trim on Flxible Starliner coaches varied. Some had full stainless steel paneling on the sides, while others had fully paintable panels. This Starliner operated by S.M.B Stage Lines of Des Moines, Iowa, boasted stainless steel on a portion of its sides. S.M.B Stage Lines had an extensive route network in Missouri, but finished its days with a Des Moines-Sioux City, Iowa route. The full company name Sedalia-Marshall-Boonville Stage Line was used when operating in Missouri. It is interesting to note that the company went into the air taxi business in 1967, with its aircraft carrying the S.M.B Stage Line name.

This Flxible Starliner, Model 218DD1-6329IC, was added to the fleet of Boise, Idaho-based Northwestern Stage Lines in 1963. The Starliner had a General Motors 4-71 diesel engine as standard equipment. About 275 Starliners were built, and it was the last of the Clipper-style body that originated in the late 1930s. Northwestern Stage Lines began in the early 1920s and operated a route between Boise and Lewiston, Idaho. Now a member of the Trailways organization, the company continues today with expanded service to Spokane, Seattle and Tacoma, Washington.

# Flxible Vistaliner Coaches

In 1955 The Flxible Company designed an entirely new coach known as the Vistaliner VL-100 with a model designation of 228JT1. It had a Cummins JT 600, 401-cubic-inch diesel engine. Seating was for 37 passengers on two levels. Baggage storage was under the floor. Continental Trailways companies purchased 126 of the 208 Vistaliners built between 1954 and 1958. One of the Continental Trailways' Vistaliners is pictured here.

# Flxible Hi-level Coaches

The Flxible Hi-level coach was introduced in 1959. It was similar to the Vistaliner but did not have a split-level seating arrangement. All seating was on a raised level, with the driver positioned lower. The Hi-level coach had a General Motors 6-71 diesel engine. Missouri Transit Lines of Macon, Missouri purchased this Flxible Hi-level coach in 1961. Missouri Transit Lines originally had extensive service in Missouri but, after selling some of its routes, concentrated mainly on a Cedar Rapids, Iowa-Springfield, Missouri route.

Continental Trailways acquired a number of Flxible Hi-level coaches in the late 1950s and 1960s. This Flxible Hi-level, Model 236DD1-60-39IC-AC, went into service for Continental American Trailways in 1960. It was one of 20 coaches in the order. Continental used the name Clipper Eagles for the Hi-level coaches. Powered by General Motors 6-71 diesel engines, the Hi-level coaches had air conditioning and under-floor baggage compartments.

# Flxible Flxiliner Coaches

The Flxible Company introduced the last of its intercity model coaches in 1964. Known as the Flxiliner, it is shown here at the Flxible plant with a restored 1927 Buick stretch out coach. The Flxiliner was patterned after the Vistaliner and Hi-level coach. The Vistaliner was first produced in 1955. The Flxiliner had a General Motors 6-71 diesel engine mounted in the rear.

The Flxible Flxiliner was produced for five years, ending in 1969 when the company decided to discontinue building intercity coaches. It was a Hi-level coach seating 39 to 41 passengers. It also had torsilastic suspension and was powered by a Detroit Diesel 6-71 engine. Cummins diesel engines were an option. In 1964, Lake Shore Coach Company of Columbus, Ohio acquired this Flxiliner. Lake Shore had its beginnings in 1927 and had routes along Lake Erie in northern Ohio. In 1949, Lake Shore traded some routes with Greyhound. Although the company no longer operated the lakeshore routes, it retained the Lake Shore name.

Scenic Hawkeye Stages of Decorah, Iowa purchased this Flxible Flxiliner in April 1967 for $44,059. It was a 41-passenger model with two turnaround seats, aisle seats and a radio system with seven speakers. A Detroit Diesel 6-71 in-line diesel engine powered the coach. Scenic Hawkeye Stages began in 1954 with several routes, and continues today as Hawkeye Stages, a charter-only carrier.

# Flxible and Dina Coaches in Mexico

Hi-level Flxible coach operated by Norte de Sonora pictured in 1966 in Nogales, Mexico.

Transportes Chihuahuenes Flxible Hi-level coach pictured in 1966 in Ciudad Juarez, Mexico.

Two Flxiliner-style coaches of Transportes Estrella Blanca, SA de CV pictured in 1990 in Mexico City.

Tres Estrellas de Oro Flxiliner-style Dina coach pictured in Los Mochis, Mexico in 1990. *Wilhelm Pflug*

Dina Flxiliner-style coach of Transportes del Norte pictured in 1990 at Nuevo Laredo, Mexico. *Wilhelm Pflug*

Dina Avante model, the latest design from Dina, but still based on the Flxible Flxiliner body.

# Ansair Flxible Coaches in Australia

There is an organization in Austrailia devoted to the Flxible Clippers and is called the Flxible Clipper Club of Australia, Inc. Of the 131 Ansair Flxibles built, 100 have been located. It is reported that 25 have been restored and operating and others are owned by club members awaiting restoration or used for spare parts.

In 1950, Flxible-style Clippers were built in Austrailia by Ansair, Pty., Ltd. Following the importation of the first Flxible to Ansair in 1948. The prototype and the first two Ansair-built Clippers had small windows. These were operated by Pioneer Tours, a part of the Ansair organization. *Gary Driver Collection*

Ansair built 131 Flxible-style coaches between 1950 and 1960 and most were in operation by Pioneer Tours. Pictured here is one of the later models with Visicoach long windows. All but three Flxible coaches in the fleet had the long windows. *Gary Driver Collection*

A view of the right side of one of the eleven Ansair Flxible coaches that had been lengthened by four feet between December 1960 and November 1961 to accommodate 37 passengers. *Gary Driver Collection*

Shown here is the driver's position of one of the Ansair Visiliner coaches with the driver seated on the right side and the passenger door on the left. Australia rule of the road is to the left. *Gary Driver Collection*

# Restored Flxible Coaches

Pictured is a 1930 Flxible-bodied Buick coach. It is owned by Brenda and Reg DeNure, who for many years owned the Chatham Coach Line in Chatham, Ontario. They purchased this coach from a convent in Michigan. It was used for outings and then lovingly stored for years in a barn. *Loring M. Lawrence*

This 1946 Flxible Clipper, Model 29BR46 was originally restored by the late Richard Maguire, at one time, the president of Capitol Bus Company in Harrisburg, Pennsylvania. The Museum of Bus Transportation now owns this coach.

Voigt's Bus Service, St. Cloud, Minnesota has restored this Flxible Visicoach, originally purchased by Northland Greyhound Lines (see page 89). It had an extra set of wheels in the rear, but these wheels have been removed. Voigt's Bus Service restored this Flxible in 1997 to commemorate the 50th anniversary of the company. It has 1.3 million miles on the original General Motors 4-71 diesel engine. The coach itself celebrated 50 years in 2003. *Darwin Voigt*

In Mexico City, Mexico, is this restored Flxible Visacoach, which at one time was operated by Transportes Frontera between Mexico City and Laredo, Texas.

111

Gary Driver, managing director of Driver Coaches, Mt. Waverly, Australia has restored this Ansair Flxible Visicoach. It is one of the 131 Flxible-style coaches built by Ansair Industries in Australia between 1950 and 1960. The top view shows the coach at Hamilton, Victoria in Australia posing with a Douglas DC-3 aircraft. The bottom view pictures the coach at Ballarat, Victoria. This restored coach originally had a Leyland engine, but it was re-engined with a General Motors 4-71 diesel engine. *Gary Driver*

# Time Tables Featuring Flxible Coaches

114

# Flxible Sales Literature

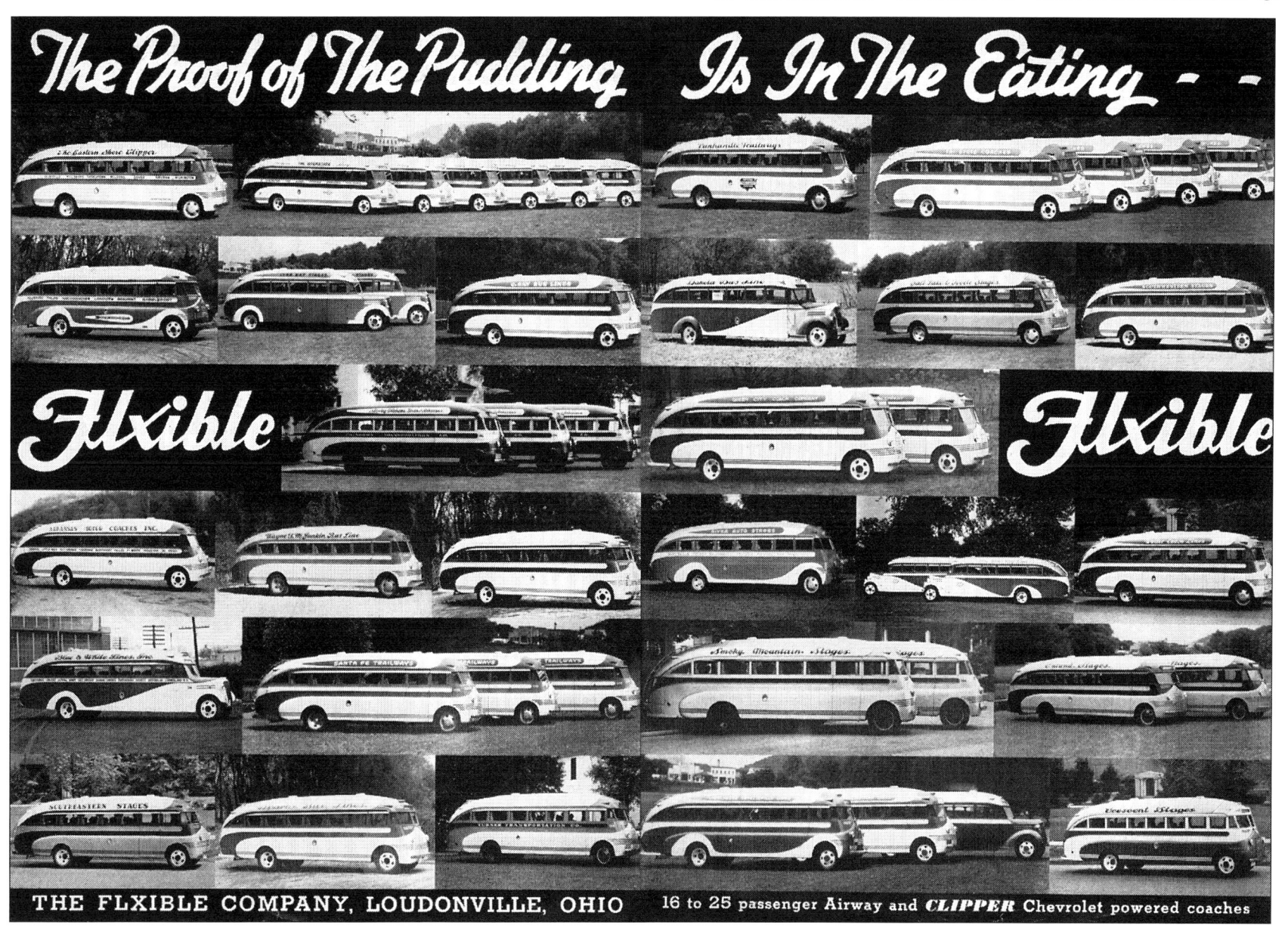

A Flxible sales brochure advertising both Airway and original Clipper models.

This Flxible sales piece described the new Flxible 25-passenger Clipper.

116

Specific features of the Flxible Clippers of 1940 were described in these advertisements.

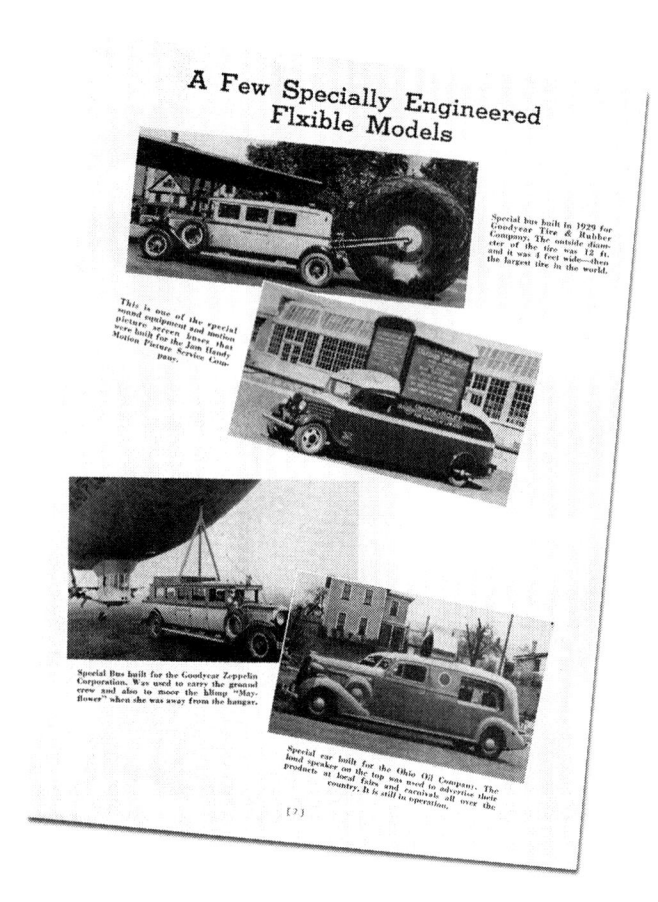

*Flxible Clipper-ings*, a monthly publication of the Flxible Company, described the history of the company as well as honored World War II achievements by plant workers.

118

The Flxible Clipper of 1948 was presented in this booklet.

Flxible Visicoaches were described with information and pictures in this detailed booklet.

120

The Flxible 37-passenger dual-engined coach was previewed in this folder.

# Flxible
## DUAL ENGINE 37-C1 COACH

## SPECIFICATIONS

| | |
|---|---|
| Overall Length | 34' 10 7/8" |
| Overall Width - Exterior | 95 1/2" |
| Overall Width - Interior | 91 15/32" |
| Overall Height - Empty | 120 1/4" |
| Headroom (aisle floor) | 76 1/4" |
| Aisle Width (between seats) | 15 1/2" |
| Headroom (seat floor to luggage racks) | 54 1/2" |
| Step Heights - ground to first | 12 5/8" |
|               - first to second | 14" |
|               - second to aisle | 12 5/8" |
| Wheelbase | 231" |
| Track - front | 80" |
|       - rear | 71 1/2" |
| Tire Size (single front, dual rear) | 10.00 x 20 |
| Curb Weight | 18,525 lbs. |
| Turning Radius (smallest) | 37' 6" |
| Seating Capacity | 37 passengers |
| Deflection Rate, rear springs | 1,200 pounds |
| Deflection Rate, front springs | 1,000 pounds |

*Specifications, materials, and details of construction are subject to correction or change without notice.*

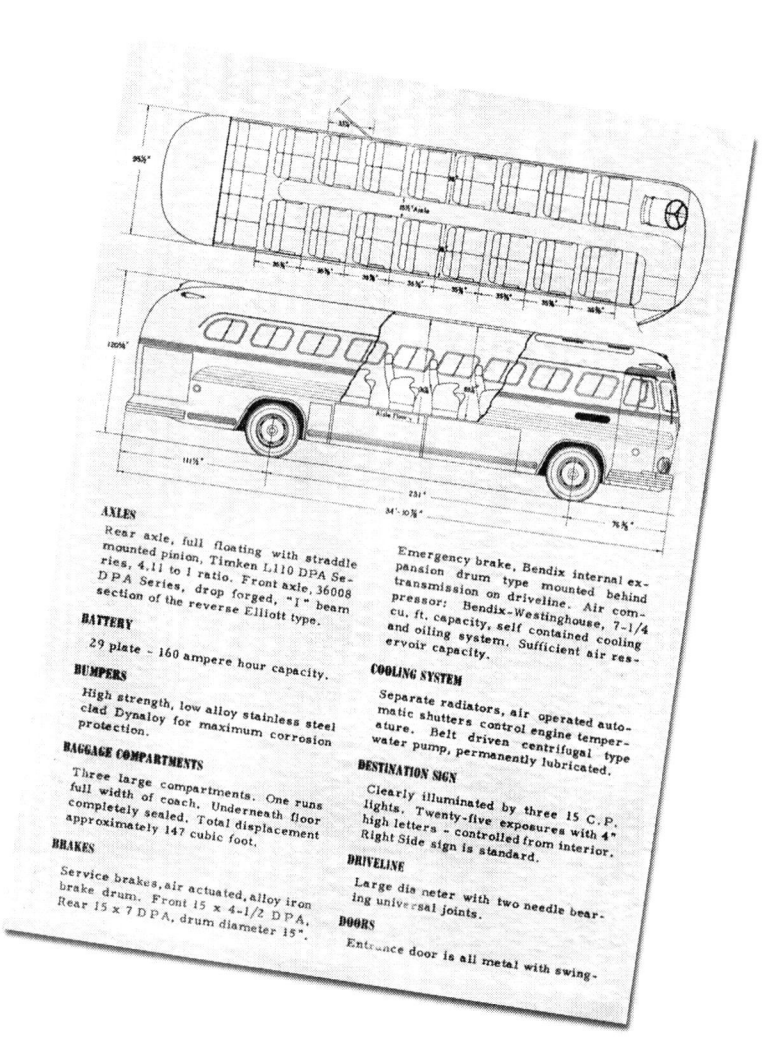

**THE FLXIBLE COMPANY ★ LOUDONVILLE, OHIO**

### AXLES
Rear axle, full floating with straddle mounted pinion, Timken L110 DPA Series, 4.11 to 1 ratio. Front axle, 36008 DPA Series, drop forged, "I" beam section of the reverse Elliott type.

### BATTERY
29 plate - 160 ampere hour capacity.

### BUMPERS
High strength, low alloy stainless steel clad Dynaloy for maximum corrosion protection.

### BAGGAGE COMPARTMENTS
Three large compartments. One runs full width of coach. Underneath floor completely sealed. Total displacement approximately 147 cubic foot.

### BRAKES
Service brakes, air actuated, alloy iron brake drum. Front 15 x 4-1/2 DPA. Rear 15 x 7 DPA, drum diameter 15".

Emergency brake, Bendix internal expansion drum type mounted behind transmission on driveline. Air compressor: Bendix-Westinghouse, 7-1/4 cu. ft. capacity, self contained cooling and oiling system. Sufficient air reservoir capacity.

### COOLING SYSTEM
Separate radiators, air operated automatic shutters control engine temperature. Belt driven centrifugal type water pump, permanently lubricated.

### DESTINATION SIGN
Clearly illuminated by three 15 C.P. lights. Twenty-five exposures with 4" high letters - controlled from interior. Right Side sign is standard.

### DRIVELINE
Large diameter with two needle bearing universal joints.

### DOORS
Entrance door is all metal with swing-

Specifications were detailed for the Flxible dual-engined 37-C1 coach in this brochure.

# Flxible
## SALES DISPLAY COACHES

The Flxible Company, long a leader in the Bus Industry, is equipped in its Special Products Division to design and build anything from a "shell" to a complete and modern Display Coach.

Past experience has indicated that some companies desire to install all or part of the interior display components - others desire a unit ready for the road. Some want considerable floor space - whereas others require but little working area.

Flxible builds coaches of practically every size to accommodate these individual requirements.

Reports from users of Sales Display Coaches show conclusive evidence that they lead to increased sales. Investigate to determine by what percentage they'll increase sales for your company.

Painted in colorful company colors and with company name and trademark prominently featured, a Flxible Sales-Display Coach fast becomes known and welcomed throughout its territory.

Sales display Flxible coaches were marketed with this interesting folder.

## SPECIFICATIONS

### INTERCITY STARLINER

**218" WHEELBASE MODEL**

| | |
|---|---|
| Overall Length | 34' 2 3/4" |
| Body Width | 95 1/2" |
| Inside Body Width | 92 1/2" |
| Overall Height | 115 9/16" |
| Overhang - Front | 75 3/4" |
| Overhang - Rear | 117" |
| Wheelbase | 218" |
| Interior Headroom at Aisle Floor Level | 78 1/2" |
| Interior Headroom at Seat Floor Level | 71 1/2" |
| Entrance Door Step Height from Ground (Empty) | 17" |
| Weight - Empty - G. M. 4-71 Power - Diesel | 16,315 lbs. |
| White 390 A S Power - Gasoline | 16,025 lbs. |

**THE FLXIBLE COMPANY • LOUDONVILLE, OHIO**

---

## Specifications

### THE Starliner
#### A NEW CONCEPT IN MOTOR COACHES
#### by Flxible

**THE FLXIBLE COMPANY • LOUDONVILLE, OHIO, U.S.A.**

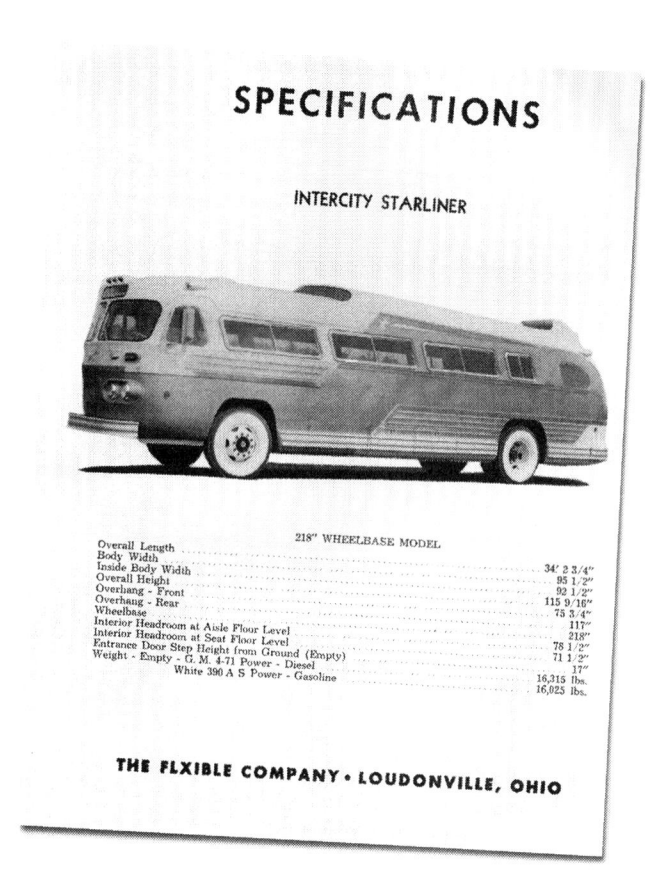

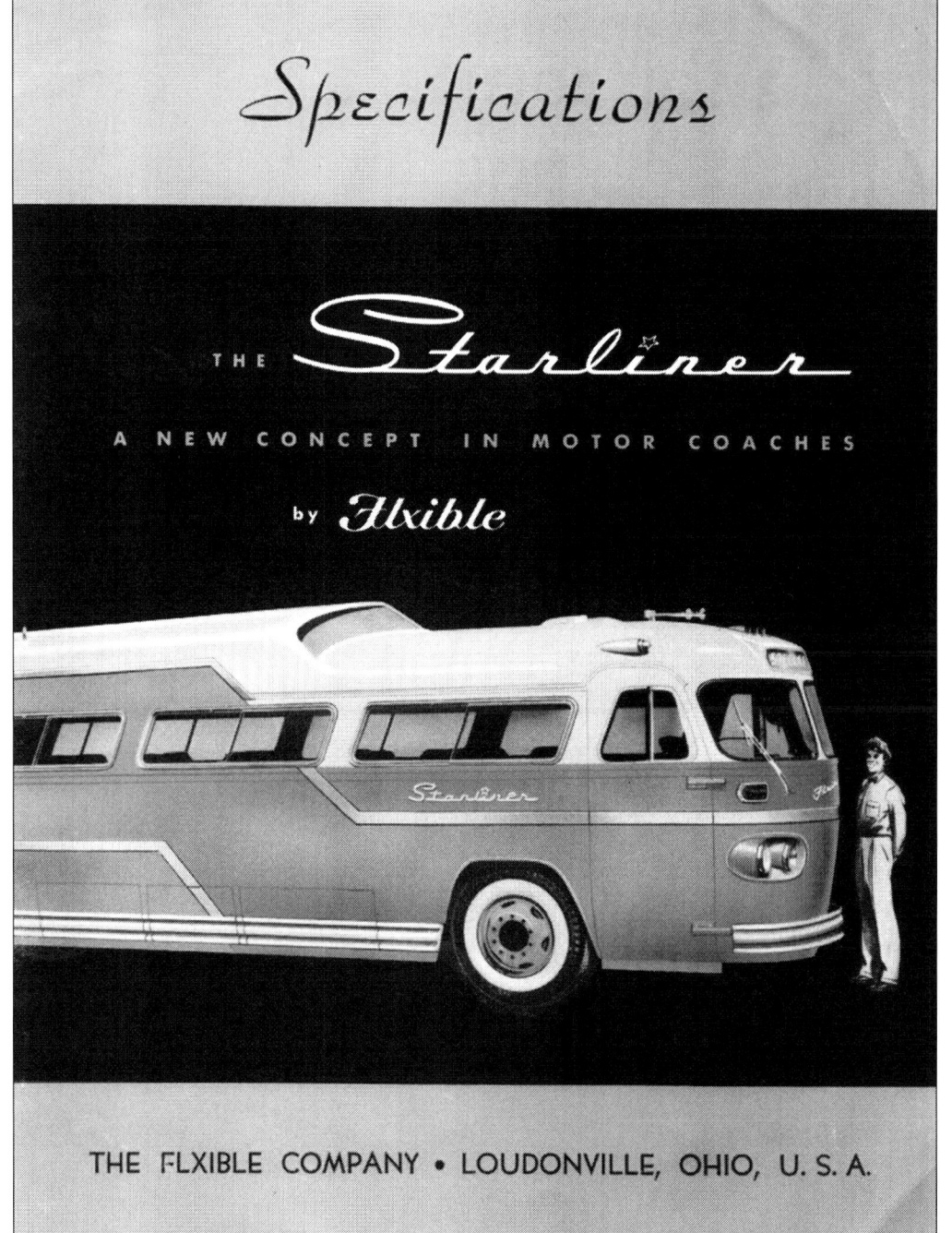

The Flxible Starliner introduced in 1955 had specifications, pictures and drawings in this four-color booklet.

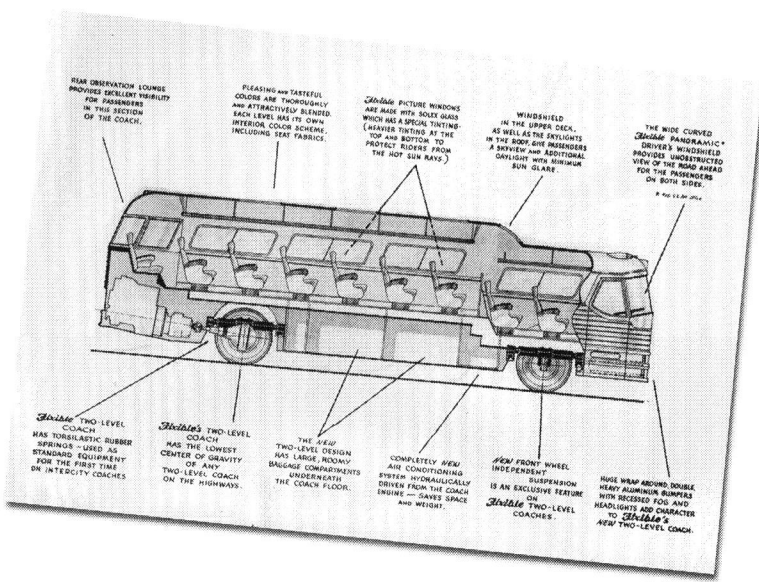

The two-level Flxible Vistaliner specifications were outlined in this detailed piece.

# A Word from the Author

Bus transportation was of interest to me when I was a student in school in Virginia, Minnesota. I spent quite a bit of time at the local bus station, learning about the bus industry. I visited with drivers and ticket agents and observed the buses when they were at the station.

There was a small intercity bus company, Northern Transportation Company, operating into Virginia. Although I had seen pictures of Flxible buses, I was able to have a good look at the Northern Flxible Airway coach on occasions at the bus station. In 1940, Northern took delivery of a rear-engined Flxible Clipper. Its styling was the ultimate for modern buses at the time, I thought. I rode the new Clipper on several occasions. The ride was excellent, and I really liked the sound of the Buick straight-eight engine and the air horns.

As I began traveling, I rode Flxible coaches on a number of occasions, especially in Minnesota, Wisconsin and the Dakotas. I traveled on the Flxibles of Zephyr Lines, Southwestern Stages, Jack Rabbit Lines, Wisconsin Northern Transportation Company, Triangle Transportation Company and others.

Then while serving in the army there were more opportunities to travel on Flxible coaches. Arkansas Motor Coach Company operated one of the Flxibles I rode. Following my army experience, I traveled quite a bit through the United States and Canada. One of the longest trips I made was on a Flxible coach from Portland, Oregon to Salt Lake City, Utah.

I had made models of Flxible coaches and a picture of them was featured in one of the Flxible Company's newsletters. The Flxible Company had me on their mailing list for photographs of Flxible coaches in sales literature for many years. On one occasion, a Flxible salesman, realizing my interest in Flxible coaches, gave me several hundred Flxible photographs and other items when he retired. This enhanced my bus transportation library.

In addition, I took pictures of Flxible coaches because they were everywhere and were interesting photo subjects. I was able to add several hundred more Flxible photographs to my library through the years.

When I entered the bus industry trade publication business, I visited many bus companies. There were many conversations about the Flxible intercity coach days. Everyone seemed to like the Flxible coaches, especially the Flxible Clippers.

The Flxible Company went into the transit bus business and I continued to have a good relationship with them. I was kept on the mailing list and pictures and information were continuously added to my bus transportation library.

Reminiscing about the bus industry's first century, there were many buses that remain as classic leaders of the industry, and certainly the Flxible Clipper is one of those buses on the top of the list. It is also good to know that there are many Flxible coaches preserved, either made into motor homes or preserved as normal passenger coaches. Most owners are members of Flxible Owners International.

The book *Bus Industry Chronicle*, which I authored, contains more information and pictures about Flxible intercity coaches.

Additional bus industry history has been recorded in several other Photo Archive books I have authored or co-authored. These books are: *Greyhound Buses 1914-2000 Photo Archive, Trailways Buses 1936-2001 Photo Archive, Yellow Coach Buses 1923-1943 Photo Archive, Trolley Buses 1913-2001 Photo Archive, Fageol & Twin Coach Buses 1922-1956 Photo Archive,* and *Prevost Buses 1924-2002 Photo Archive.*

**Bill Luke is shown standing in front of a 1949 Northern Transportation Company Flxible Clipper on a snowy winter day in Virginia, Minnesota.**